MISSOURI
TOTAL ECLIPSE GUIDE

Commemorative Official
Keepsake Guidebook

2017 Total Eclipse State Guide Series

Aaron Linsdau

SASTRUGI PRESS

JACKSON HOLE

Sastrugi Press / Published by arrangement with the author

Missouri Total Eclipse Guide: Commemorative Official Keepsake Guidebook

The author has made every effort to accurately describe the locations contained in this work. Travel to some locations in this book is hazardous. The publisher has no control over and does not assume any responsibility for author or third-party websites or their content describing these locations, how to travel there, nor how to do it safely. Refer to official forest and national park regulations.

Any person exploring these locations is personally responsible for checking local conditions prior to departure. You are responsible for your own actions and decisions. The information contained in this work is based solely on the author's research at the time of publication and may not be accurate in the future. Neither the publisher nor the author assumes any liability for anyone climbing, exploring, visiting, or traveling to the locations described in this work. Climbing is dangerous by its nature. Any person engaging in mountain climbing is responsible for learning the proper techniques. The reader assumes all risks and accepts full responsibility for injuries, including death.

Park maps are courtesy of the National Park Service.

Sastrugi Press
PO Box 1297, Jackson, WY 83001, United States
www.sastrugipress.com
Quantity sales: Special discounts are available on quantity purchases by corporations, associations, and others. For details, contact the publisher at the address above.

Library of Congress Catalog-in-Publication Data
Library of Congress Control Number: 2017905385
Linsdau, Aaron
Missouri Total Eclipse Guide / Aaron Linsdau-1st United States edition
p. cm.
1. Nature 2. Astronomy 3. Travel 4. Photography
Summary: Learn everything you need to know about viewing, experiencing, and photographing the total eclipse in Missouri on August 21, 2017.

ISBN-13: 978-1-944986-10-0
ISBN-10: 1-944986-10-3

508.4—dc23

Printed in the United States of America
All photography, maps and artwork by the author, except as noted.

10 9 8 7 6 5 4 3 2 1

Contents

Introduction

Thank you for purchasing this book. It has everything you need to know about the total eclipse in Missouri on August 21, 2017.

A total eclipse passing across the United States is a rare event. The last US total eclipse was in 1979. It traveled over Washington, Oregon, Montana, and the corner of North Dakota.

The next total eclipse over the US will not be until April 8, 2024. It will pass over Texas, the Midwest, and on to Maine. After that, the next coast-to-coast total eclipse will be in 2045!

It's imperative to make travel plans today. You will be amazed at the number of people swarming to the total eclipse path. Some might say watching a partial versus a total eclipse is a similar experience. It's not.

This book is written for Missouri visitors and anyone else viewing the eclipse. You will find general planning, viewing, and photography information inside. Should you travel to the eclipse path in Missouri in mid-August, be prepared for an epic trip. The state estimates a half million visitors will converge on Missouri.

Most hotels in the communities and cities along the path of totality in Missouri have already sold out as of the writing of this guide. Finding lodging along the eclipse path will be a major challenge.

Resources will be stretched far beyond the normal limits. Think gas lines from the late 1970s. It may be likely that traffic along mountain highways will come to a complete standstill during this event. Be prepared with backup supplies.

Many smaller Missouri towns are far from any major city. Missouri country roads are slow. Please obey posted speed limits within all forest and park areas. Be cautious about believing a map application's estimate of travel time in Missouri.

People in all communities along the path of the total eclipse plan to rent out their properties for this event. With a major celestial event in the summer of 2017, be assured that Missouri "hasn't seen anything yet."

Is this to say to avoid Missouri or other areas during the eclipse? Not at all! This guidebook provides ideas for interesting, alternative, and memorable locations to see the eclipse. It will be too late to rush to a better spot once the eclipse begins. Law enforcement will be out to help drivers reconsider speeding.

Please be patient and careful. There will be a large rush of people from all over the world converging on Missouri to enjoy the total eclipse. Be mindful of other drivers on eclipse weekend, as they may not be familiar with Missouri roads.

You should feel compelled to play hooky on August 21. Ask for the day off. Take your kids out of school. They'll likely be adults before the next chance to see a total eclipse. Create family memories that will last a lifetime. Sastrugi Press does not normally advocate skipping school or work. Make an exception because this is too big an event to miss.

Wherever you plan to be along the total eclipse path, leave early and remember your eclipse glasses. People from all around the planet will converge on Missouri. Be good to your fellow humans and be safe. We all want to enjoy this spectacular show.

Visit www.sastrugipress.com/eclipse for the latest updates for this state eclipse book series.

AUTHOR INFORMATION

Polar explorer and motivational speaker Aaron Linsdau's first book, *Antarctic Tears*, is an emotional journey into the heart of Antarctica. This Jackson Hole native ate two sticks of butter every day to survive. Aaron coughed up blood early in the expedition and struggled through equipment failures. Despite the endless difficulties, he set a world record for surviving the longest solo expedition to the South Pole.

Aaron teaches audiences how the common person can achieve uncommon results. He shares his techniques for overcoming adrenaline burnout and constant overload. He inspires audiences to face their challenges with a new perspective. Aaron builds grit, teaches courage, and shows how to maintain a positive attitude in the face of adversity. He hopes that you will be inspired and have an enjoyable time watching the total eclipse in Missouri.

Visit his websites: www.aaronlinsdau.com and www.ncexped.com

All About Missouri

OVERVIEW OF MISSOURI

Missouri, the Show-Me State, is a wonderful place to visit in the summertime, as it has a diverse selection of things to do. Between the forests, museums, historical national monuments, and theme parks, there is something to appeal to every member of the family. Both children and adults will fall in love with the unique scenery, history, and friendly people of the great state of Missouri. Missouri promises to be an excellent state to visit before, during, and after the total eclipse.

Perhaps the most famous historical landmark that Missouri has to offer is the Gateway Arch. It was built to commemorate the role that Thomas Jefferson and the city of St. Louis played in the westward expansion that helped make America what it is today. Standing sixty-three

stories tall, the Gateway Arch is impressive to see in person and should be experienced by all. There is a tram that visitors can take to the top of the arch. The view of St. Louis from this vantage point gives you the feeling of flying like a bird, high above the city.

If you want something with a little more glitz and glamour, then the city of Branson is for you. It is home to several theaters, amusement parks, and museums. Branson is sure to entertain your entire family. It is home to the Dixie Stampede show, an entertainment extravaganza featuring musical performances, jumping horses, and excellent special effects. This is all done before your very eyes while you feast on a home-cooked meal.

Branson is also home to the legendary Shoji Tabuchi Theatre, where violin virtuoso Shoji Tabuchi performs to captivated audiences most nights out of the week. His multimedia stage shows have wowed audiences for years, and your whole family will enjoy it.

The Mississippi River runs right through Missouri and provides lots of fun summer vacation activities. It's a world-renowned destination for people who love to fish. The Mississippi River Road follows the river down the state and makes for a truly stunning drive. You can take a relaxing trip driving down the road and just absorb the natural majesty of that famous river.

Farther down the river, in the city of Hannibal, you can visit the boyhood home of famous author and orator Samuel Clemens, better known around the world as Mark Twain. Along with his boyhood home is a museum erected in his honor. There are also five dams along the river in Missouri, one of them in Canton, and they make for some very pretty sights.

The Mark Twain National Forest is a gorgeous wilderness to explore. It's a huge 1.5 million acres and completely open to the public. There are over 750 miles of trails that can be used for hiking, horse riding, biking, or driving. It has plenty of campgrounds and 350 miles of streams that you can lazily paddle a canoe down or ride a kayak. The Mark Twain National Forest is one of the most beautiful national forests in all of the United States and is an absolute must-see for anyone traveling to Missouri.

For those interested in the history of steamboats, a visit to the Steamboat Arabia Museum is in order. Boasting the largest single collection of pre-civil war artifacts, the Steamboat Arabia museum is one of the must-see museums in Kansas City. All the artifacts in the museum had been recovered from a steamboat that had sunk in the dangerous Missouri River.

Over sixty tons of artifacts are still being cleaned and cataloged by the preservationists at the museum, so with every repeat visit you make, there are bound to be more and more amazing historical artifacts that are new to the collection. Seeing those actual pieces of pre-civil war American history puts visitors in touch with a piece of the past that people do not normally get to be a part of. It is sure to create memories that won't be soon forgotten. This is the kind of experience that you can only have in Missouri.

Missouri is bound to entertain your whole family and will surely lure you back for many repeat visits. With so many forests, parks, museums, monuments, and amusement parks, you'll have to make several trips before you have seen everything. A trip to Missouri can be relaxing, educational, revitalizing, and most of all it will be a whole lot of fun for the entire family.

Hotels and Motels During the Eclipse

Once word of the total eclipse over Missouri spreads, rooms will become scarce. Many hotels in towns along the path of totality in western states have been sold out for a year or more. Missouri is not alone in this challenge. Hotels in Idaho and Wyoming have been sold out for multiple years in anticipation of the total eclipse.

What does this mean for eclipse visitors? Lodging and room rentals in eclipse towns will be at a massive premium. Does that mean all hope is lost to find a place to stay? Not at all. But you will have to be creative. There will be few if any hotel rooms available in these eclipse cities by the time this book is printed. Accommodations in the cities and towns along the path of the eclipse have been sold out for months.

In spring 2017, the author searched on Hotels.com for rooms along the total eclipse path on the weekend of August 21 and found options still available. Once word of the eclipse spreads, room rates will increase and availability will drop.

Search for rooms farther away from the eclipse path. If you are willing to stay in cities outside the eclipse path, you will have better success at finding rooms. As the eclipse approaches, people will book rooms farther from the totality path. By midsummer, rooms in cities like Columbia may be unavailable. The effect of this event will be felt across Missouri and the rest of the United States.

Think regionally when looking for rooms. Be prepared to search far and wide during this major event. If a five-hour drive is manageable, your lodging options greatly expand, but it also increases your travel risk.

Internet Rentals

To find rooms to stay in towns along the eclipse path, try a web service such as Airbnb.com. Note that many people rent out rooms or homes illegally, against zoning regulations. Many cities have already beginning to feel the crunch of eclipse inquiries.

If Missouri towns fully enforce zoning laws, authorities may prevent your weekend home rental. Online home rentals during the eclipse will be a target for rental scams. People from out of the area steal photos and descriptions, then post the home for rent. You send your check or wire money to a "rental agent" then show up to find you have been scammed. If the deal sounds strange or too good to be true, run away.

CAMPING

If you can book a campsite, do it now. Do not wait. All areas in the national forests are first-come, first-served. Forest roads will be packed. Expect all areas to be swarming with people. Show up early to stake out your spot. Consider staying farther away and driving early on August 21.

Please respect private land too. Missouri folks don't take kindly to people overrunning their property without permission. In a big state with six million residents, people are very protective, but they're friendly, too. You never know what you might be able to arrange with a smile and a bit of money.

This all said, there are plenty of camping opportunities throughout Missouri. You don't have to sleep exactly on the eclipse path. If you're ready to rough it, there are national forest camping options.

Government agencies have been meeting since 2015 to talk about how to manage the influx of people. Every possible government agency will be working full time to enforce the various rules and regulations.

NATIONAL PARKS AND MONUMENTS

Chances are finding a camping site at any state park, national park, or national monument in Missouri will be challenging. To watch the eclipse from any location, you do not have to sleep in it. You just need to drive there in the morning.

Law enforcement will be present on the eclipse weekend. Hundreds of thousands of people are expected in the region. Parking will overflow. It will make parking lots and lines on Black Friday at the mall look uncrowded. For an event of this magnitude, you'll need to find your parking space early.

The first sentence of the national parks mission statement is:

> *"The National Park Service preserves unimpaired the natural and cultural resources and values of the national park system for the enjoyment, education, and inspiration of this and future generations."*

Roadside camping (sleeping in your car) is not allowed in national monuments or parks. Park facilities are only designed to handle so many people per day. Water, trash collection, and toilets can only withstand so much. If you notice trash on the ground, take a moment to throw it away. Protect your national park and help out. Rangers are diligent and hardworking but they can only do so much to manage the expected crowds.

National Forests and Wilderness

There are national forest options in Missouri. They all have camping opportunities. The forest service manages undeveloped and primitive campsites. Be sure to check for any fire restrictions. Check with individual agencies for last-minute information and regulations. The forest service requires proper food storage. Plan to purchase food and water before choosing your campsite. Below is a partial list of national forests along or near the total eclipse path:

Mark Twain NF: www.fs.usda.gov/main/mtnf
Bell Mountain Wilderness
Rockpile Mountain Wilderness
Mingo National Wildlife Refuge and Wilderness

Backcountry service roads abound in Missouri. Maps for forests are available at local visitor centers and bookstores. This book's website has digital copies of some forest maps.

Printed national forest maps are large and detailed. They have illustrated road paths, connections, and other vital travel information not available on digital device maps. Viewing digital maps on your smartphone or iPad is difficult. If you plan to camp in the forest, a real paper map is a wise investment.

Camping in federal wilderness areas is also allowed. Those areas afford the ultimate backcountry experience. However, be aware that no vehicle travel is allowed in the specially designated areas. This ban includes: vehicles, bikes, hang gliders, and drones. You can travel only on foot or with pack animals.

Sleep in Your Car

Countless RVs, campers, trucks, cars, and motorcycles will flood Missouri. Sleeping in your car with friends is tolerable. Doing so with unadventurous spouses or children is another matter.

Do not be caught along the path of the total eclipse without some sort of plan, especially in the bigger cities of Missouri. The whole path of totality will fill with people on August 21. It will be anything but normal.

Useful Local Webcams

Local webcams are handy to make last-minute travel decisions. The webcams are sensitive enough to show headlights at night. Use them to determine if there are issues before traveling out. Eclipse traffic will add to the morning commuter traffic.

The smartphone application Wunderground is useful to check on webcams in one place. All the webcams are listed in the app.

Weather

It's all about the weather during the eclipse. Nothing else will matter if the sky is cloudy. You can be nearly anywhere in Missouri and catch a view of the sky when traffic comes to a standstill. But if there's a cloud cover forecast, seriously reconsider your viewing location.

Travel early wherever you plan to go. Attempting to change locations an hour before the eclipse due to weather will likely cause you to miss the event. Missouri country roads can be narrow and slow. The number of vehicles will cause unexpected backups.

MODERN FORECASTS

Use a smartphone application to check the up-to-date weather. Wunderground is a good application and has relatively reliable forecasts for the region. The hourly forecast for the same day has been rather accurate for the last two years. The below discussion refers to features found in the Wunderground app. However, any application with detailed weather views will improve your eclipse forecasting skills.

CLOUD COVER FORECAST

The most useful forecast view is the visible and infrared cloud-coverage map. Avoid downloading this app the night before and trying to learn how to read it. Practice reading them at home. It's imperative to understand how to interpret the maps early.

Infrared cloud map showing the worst case eclipse cloud cover.
Courtesy of National Weather Service.

All cloud cover, night or day, will appear on an infrared map. Warm, low-altitude clouds are shown in white and gray. High-altitude cold clouds are displayed in shades of green, yellow, red, and purple. Anything other than a clear map spells eclipse-viewing problems.

To improve your weather guess, use the animated viewer of the cloud cover. It will give you a sense of cloud motion. You can discern whether clouds or rain are moving toward, away from, or circulating around your location.

NORMAL MISSOURI WEATHER PATTERN

Due to the direction of the jet stream, most weather travels across the Pacific Ocean, through the western states, over Kansas and Nebraska, and then into Missouri. On occasion, weather can approach from any direction. Due to the nature of the tropical storms from the

Gulf of Mexico, weather in Missouri can be unpredictable.

The common weather pattern in August is hot and humid. Passing cold fronts in summer can bring unexpected cloud cover.

Cities in Missouri tend to have variable clouds during August. Prepare to make adjustments. If anything other than clear skies are predicted, plan to drive to other parts of Missouri, Nebraska, Kansas, Illinois, Kentucky, or Tennessee.

Be aware of tornadoes in Missouri. Although the peak tornado season is June, there have been many recorded tornadoes in August. Pay attention to the weather forecast. If dangerous weather is predicted, your main concern should be safety rather than chasing an eclipse.

Consider that slow-moving clouds can obscure the sun for far longer than the two-minute duration of the totality. The time of totality is so short that you do not want to risk it. Missing it due to a single cloud will be a major disappointment.

Local Eclipse Weather Forecasts

Local town and city newspapers, radio, and television stations around Missouri will have a weekend edition with articles discussing the eclipse weather. However, conditions change unpredictably in Missouri. A three-day forecast may be completely incorrect.

Forest Fires

For the past several years, forest fires have been common in the United States. The summer of 2017 is likely to be no different. There were large fires in the forests Missouri during 2015. Chances are there will be fires again in the region in the summer of 2017. For fire updates check:

inciweb.nwcg.gov.

For the best eclipse viewing experience, you need to have as clear a sky as possible. Fog, clouds, or smoke will obscure the subtleties of the sun's corona. If you think the view of the sky is going to be blocked, don't wait until the last minute to move to a clearer location. If you wait too long to decide to move to a better viewing area, it may be impossible due to traffic.

Road Closures Due to Fires

Highways connecting various Missouri towns can be closed during a major fire. The Reynolds County fire in 2015 disrupted travel in the area and blackened the sky in the Mark Twain National Forest.

With unpredictable weather in the last few years, it's a guess what will happen in August 2017. If forest areas continue to remain wet, the whole region may dodge the fires of years previous.

Although fire is an important part of forest ecology, it worries eclipse chasers. Other than clouds, smoke from fires will block the view of the sun and moon on the morning of the eclipse. Should there be fires where you are or may be headed, reconsider your location as early as possible. The most accurate website for fires is:

inciweb.nwcg.gov

Check the Missouri road report for updated closure information:

traveler.modot.org/

It's imperative to plan for fires and their effects. Watch the weather reports. If strong winds and lightning storms are forecast, prepare to change your viewing location. If conditions are poor, you and thousands of other vehicles will be trapped in slow-moving traffic.

If you believe it's necessary to leave a town to watch the eclipse, do so the night before or extremely early in the morning. RVs are common, and trains of them crawl through popular areas.

Missouri Information

Cellular Phones

Cellular "cell" phone service in remote Missouri locations is spotty at best. Most of the time there is good coverage along the main highways and interstates. However, even along major thoroughfares, there can be little or no coverage.

It's possible to find zones where text messages will send when phone calls are impossible. If you cannot make a phone call, the chance of having data coverage for web surfing or e-mail is low.

Please look up any information or communicate what you need before departing from the main roads around Missouri. Bureau of Land Management (BLM) areas sometimes have coverage. Planned to be self-contained. Treat like your cell phone like it won't connect.

You may find yourself out of cell service. With a large number of cell users in a concentrated area, coverage and data speed may collapse as well. Search on the phrase "cell phone coverage breathing".

Wilderness and Forest Safety

All Missouri forests are full of wild animals. Although beautiful, wild animals can be dangerous. They can easily injure or kill people, as they are far more powerful than humans. Do not try to feed any wild animals, including squirrels, foxes, and chipmunks, as they can carry diseases. These suggestions apply to all public lands.

Bear

The forests of Missouri are home to black bears. Safety is imperative around these powerful animals. Although they often appear docile, they can become aggressive if threatened. In the unlikely event of an attack, fight back against the bear. Use whatever you have at your disposal to defend yourself. Report all negative encounters or aggressive bears to the local authorities.

If a bear hears you, it will usually vacate the area. Bear charges are often caused by unexpected and surprise encounters. Noise is the best defense to avoid surprising bears. Regularly clap, make noise, and talk loudly. The Missouri Department Conservation Bear Safety website has more specific information on safety and food management in bear country: nature.mdc.mo.gov/discover-nature/general-species-information/be-bear-aware

It is recommended to stay one hundred yards (300 feet) away from all bears. They are exciting to see but need their space. Refer to current forest or park regulations for more safety information.

RATTLESNAKES

There are several primary species of rattlesnakes in Missouri: Timber Rattlesnake, Massasauga, Osage Copperhead, Western Copperhead, and Western Pygmy Rattlesnake. Although these reptiles are not generally aggressive, they can strike when provoked or threatened. Of the approximately 8,000 people annually bitten by venomous snakes in the United States, ten to fifteen people die according to the U.S. Food and Drug Administration.

The best way to avoid rattlesnake encounters is to be mindful of your environment. Do not place your hands or feet in locations where you cannot clearly see the surroundings. Avoid heavy brush or tall weeds where snakes hide during the day. Step on a log or rock rather than over it, as a hidden snake might be on the other side. Rattlesnakes may not make any noise before striking.

Avoid handling all rattlesnakes. Should you be bitten, stay calm and call 911 or emergency dispatch as soon as possible. Transport the victim to the nearest medical facility immediately. Rapid professional treatment is the best way to manage rattlesnake bites. Refer to US Forest Service and professional medical texts for more information on managing rattlesnakes injuries.

TICKS

Ticks exist all across the United States but not all species transmit disease. Ticks cannot fly or jump, but they climb grasses in shrubs in order to attach to people or animals that pass by. Ticks feed on the blood of their host. In doing so, they can transmit potentially life-threatening diseases such as Lyme disease.

According to the National Pesticide Information Center, ticks must be attached and feed for several hours before an infection can be passed. Do not wait that long to manage tick exposure. As soon as you travel through outdoor locations, especially in grasses or shrubs, check yourself for ticks. They can be incredibly small and difficult to detect. Some are smaller than the head of a pin.

You may wish to consider wearing tick-specific insect repellent. Check with your doctor or medical professional about any potential adverse side effects of chemical repellents.

Should you discover an attached tick, follow modern tick removal methods. There are many incorrect tick removal methods found on-line. Refer to your medical professional for proper removal methods. Improper removal increases the likelihood of an infection.

SCORPIONS

Scorpions are generally found in the dry areas of Missouri. However, they may be anywhere in the state. Avoid all contact with these arthropods, as their sting is painful and dangerous. For more information, visit the CDC webpage: www.cdc.gov/niosh/topics/insects.

Eclipse Day Safety

1. Hydrate

Summers can be extremely warm. The excitement of the event can distract you from managing hydration. Drink plenty of water. Consume more than you would at home.

2. Eye Safety time

Use certified eclipse safety glasses at all times when viewing the partial eclipse. Only remove the glasses when the totality happens. Give your eyes time to rest. Even though the glasses are safe, your eyes can dry out and become irritated. Bring FDA approved eye drops to keep your eyes moist.

3. Sun exposure

The sun is much more intense in August. Wear sunglasses and liberally apply sunscreen to avoid sunburns.

4. Eat well

Keep your energy up. Appetite loss is common when traveling. Maintain your normal eating schedule.

5. Prepare for temperature changes

Temperatures will drop rapidly once the sun sets, especially in the mountains or desert. Bring appropriate clothing.

6. Talk with your doctor

If the humidity, travel, or stress affects you talk with your doctor before traveling. Seek professional medical attention if you develop serious symptoms.

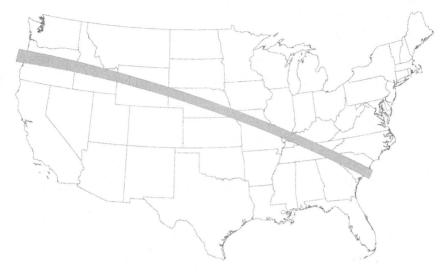

Total eclipse path across the United States (approximate).

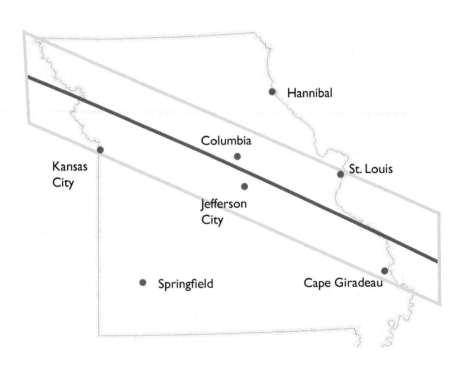

Total eclipse path across Missouri (approximate).

All About Eclipses

How an Eclipse Happens

An eclipse occurs when one celestial body falls in line with another, thus obscuring the sun from view. This occurs much more often than you'd think, considering how many bodies there are in the solar system. For instance, there are over 150 moons in the solar system. On Earth, we have two primary celestial bodies: the sun and the moon. The entire solar system is constantly in motion, with planets orbiting the sun and moons orbiting the planets. These celestial bodies often come into alignment. When these alignments cause the sun to be blocked, it is called an eclipse.

For an eclipse to occur, the sun, Earth, and moon must be in alignment. There are two types of eclipses: solar and lunar. A solar eclipse occurs when the moon obscures the sun. A lunar eclipse occurs when the moon passes through Earth's shadow. Solar eclipses are much more common, as we experience an average of 240 solar eclipses a century compared to an average of 150 lunar eclipses. Despite this, we are more likely to see a lunar eclipse than a solar eclipse. This is due to the visibility of each.

For a solar eclipse to be visible, you have to be in the moon's shadow. The problem with viewing a total eclipse is that the moon casts a small shadow over the world at any given time. You have to be in

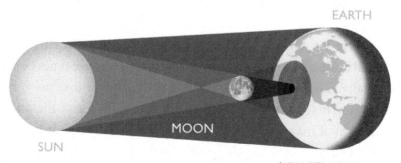

EARTH

MOON

SUN

* ILLUSTRATION
NOT TO SCALE

a precise location to view a total eclipse. The issue that arises is that most of these locations are inaccessible to most people. Though many would like to see a total solar eclipse, most aren't about to set sail for the middle of the Pacific Ocean. In fact, a solar eclipse is visible in the same place on the world on average every 375 years. This means that if you miss a solar eclipse above your hometown, you're not going to see another one unless you travel or move.

It's much easier to catch a glimpse of a lunar eclipse, even though they occur at a much lower frequency than their solar counterparts. A lunar eclipse darkens the moon for a few hours. This is different than a new moon when it faces away from the sun. During these eclipses, the moon fades and becomes nearly invisible.

Another result of a lunar eclipse is a blood moon. Earth's atmosphere bends a small amount of sunlight onto the moon turning it orange-red. The blood moon is caused by the dawn or dusk light being refracted onto the moon during an eclipse.

Lunar eclipses are much easier to see. Even when the moon is in the shadow of Earth, it's still visible throughout the world because of how much smaller it is than Earth.

TOTAL VS. PARTIAL ECLIPSE

What is the difference between a partial and total eclipse? A total eclipse of either the sun or the moon will occur only when the sun, Earth, and the moon are aligned in a perfectly straight line. This ensures that either the sun or the moon is partially or completely obscured.

In contrast, a partial eclipse occurs when the alignment of the three celestial bodies is not in a perfectly straight line. These types of eclipses usually result in only a part of either the sun or the moon being obscured. This is often what led to ancient civilizations believing that some form of magical beast or deity was eating the sun or the moon. It appears as though something has taken a bite out of either the sun or the moon during a partial eclipse.

Total eclipses, rarer than partial eclipses, still occur quite often. It's more difficult for people to be in a position to experience such an event firsthand. Total solar eclipses can only be viewed from a small portion of the world that falls into the darkest part of the moon's shadow. Often this happens in the middle of the ocean.

The Moon's Shadow

The moon's shadow is divided into two parts: the umbra and the penumbra. The former is much smaller than the latter, as the umbra is the innermost and darkest part of the shadow. The umbra is thus the central point of the moon's shadow, meaning that it is extremely small in comparison to the entire shadow. For a total solar eclipse to be visible, you need to be directly beneath the umbra of the moon's shadow. This is because that is the only point at which the moon completely blocks the view of the sun.

In contrast, the penumbra is the region of the moon's shadow in which only a portion of the light cast by the sun is obscured. When

Total eclipse shadow 2016 as seen from 1 million miles on the Deep Space Climate Observatory satellite. Courtesy of NASA.

standing in the penumbra, you are viewing the eclipse at an angle. In the penumbra, the moon does not completely block the sun from view. This means that while the event is a total solar eclipse, you'll only see a partial eclipse. The umbra for the August 21st eclipse is approximately sixty miles wide. The penumbra will cover much of the United States.

To provide some context, the last total solar eclipse we experienced occurred on March 9, 2016, and was visible as a partial eclipse across most of the Pacific Ocean, parts of Asia, and Australia. However, the only place in the world to view this total solar eclipse was in a few parts of Indonesia.

Due to the varied locations and the brief periods for which they're visible, it's difficult to see each and every eclipse that occurs. Many people don't even realize that they have occurred. Consider that the umbra of the moon represents such a small fraction of the entire shadow and the majority of our planet is comprised of water. Thus, the rarity of being able to view a total solar eclipse increases significantly because it's likely that the umbra will fall over some part of the ocean rather than a populated landmass.

ECLIPSES THROUGHOUT HISTORY

Ancient peoples believed eclipses were from the wrath of angry gods, portents of doom and misfortune, or wars between celestial beings. Eclipses have played many roles in cultures, creating myths since the dawn of time. Both solar and lunar eclipses affected societies worldwide. Inspiring fear, curiosity, and the creation of legends, eclipses have cast a long shadow in the collective unconscious of humanity throughout history.

EARLY MYTH & ASTRONOMY

Documented observations of solar eclipses have been found as far back in history as ancient Egyptian and Chinese records. Time-keeping was important to ancient Chinese cultures. Astronomical observations were an integral factor in the Chinese calendar. The first observation of a solar eclipse is found in Chinese records from over 4,000 years ago. Evidence suggests that ancient Egyptian observations may predate those archaic writings.

Many ancient societies, including Roman, Greek and Chinese civilizations, were able to infer and foresee solar eclipses from astronomical data. The sudden and unpredictable nature of solar eclipses had a stressful and intimidating effect on many societies that lacked the scientific insight to accurately predict astronomical events. Relying on the sun for their agricultural livelihood, those societies interpreted solar eclipses as world-threatening disasters.

In ancient Vietnam, solar eclipses were explained as a giant frog eating the sun. The peasantry of ancient Greece believed that an eclipse was the sign of a furious godhead, presenting an omen of wrathful retribution in the form of natural disasters. Other cultures were less speculative in their investigations. The Chinese Song Dynasty scientist Shen Kuo proved the spherical nature of the Earth and heavenly bodies through scientific insight gained by the study of eclipses.

THE ECLIPSE IN NATIVE AMERICAN MYTHOLOGY

Eclipses have played a significant role in the history of the United States. Before Europeans settled in the Americas, solar eclipses were important astronomical events to Native American cultures. In most native cultures, an eclipse was a particularly bad omen. Both the sun and the moon were regarded as sacred. Viewing an eclipse, or even being outside for the duration of the event, was considered highly taboo by the Navajo culture. During an eclipse, men and women would simply avert their eyes from the sky, acting as though it was not happening.

The Choctaw people had a unique story to explain solar eclipses. Considering the event as the mischievous actions of a black squirrel and its attempt to eat the sun, the Choctaw people would do their best to scare away the cosmic squirrel by making as much noise as possible until the end of the event, at which point cognitive bias would cause them to believe they'd once again averted disaster on an interplanetary scale.

CONTEMPORARY AMERICAN SOLAR PHENOMENA

The investigation of solar phenomena in twentieth-century American history had a similarly profound effect on the people of the United States. A total solar eclipse occurring on the sixteenth of June, 1806, engulfed the entire country. It started near modern-day Arizona. It passed across the Midwest, over Ohio, Pennsylvania, New York, Massachusetts, and Connecticut. The 1806 total eclipse was notable for being one of the first publicly advertised solar events. The public was informed beforehand of the astronomical curiosity through a pamphlet written by Andrew Newell entitled *Darkness at Noon, or the Great Solar Eclipse.*

This pamphlet described local circumstances and went into great detail explaining the true nature of the phenomenon, dispelling myth and superstition, and even giving questionable advice on the best methods of viewing the sun during the event. Replete with a short historical record of eclipses through the ages, the *Darkness at Noon* pamphlet is one of the first examples of an attempt to capitalize on the mysterious nature of solar eclipses.

Another notable American solar eclipse occurred on June 8, 1918. Passing over the United States from Washington to Florida, the eclipse was accurately predicted by the U.S. Naval Observatory and heavily documented in the newspapers of the day. Howard Russell Butler, painter and founder of the American Fine Arts Society, painted the eclipse from the U.S. Naval Observatory, immortalizing the event in *The Oregon Eclipse.*

Four more total solar eclipses occurred over the United States in the years 1923, 1925, 1932, and 1954, with another occurring in 1959. The October 2, 1959, solar eclipse began over Boston, Massachusetts. It was a sunrise event that was unviewable from the ground level. Eminent astronomer Jay Pasachoff attributed this event to sparking his interest in the study of astronomy. Studying under Professor Donald Menzel of Williams College, Pasachoff was able to view the event from an airline hired by his professor.

To this day, many myths surround the eclipse. In India, some local customs require fasting. In eastern Africa, eclipses are seen as a danger to pregnant women and young children. Despite the mystery and legend associated with unique and rare astronomical events, eclipses continue to be awe-inspiring. Even in the modern day, eclipses draw out reverential respect for the inexorable passing of celestial bodies. They are a reminder of the intimate relationship between the denizens of Earth and the universe at large.

Present Day Eclipses

The year 2016 brought the world just two solar eclipses. A total solar eclipse occurred on the 9th of March. An annular solar eclipse, in which the sun appears as a "ring of fire" occurred on the 23rd of March. If you're interested in seeing this rare and exciting solar

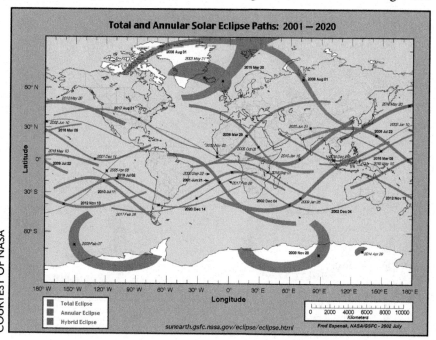

COURTESY OF NASA

phenomenon yourself, you must to travel to either South America or Western Africa on the 26th of February, 2017.

The next total solar eclipse viewable from the United States, or anywhere else in the world, will occur on the 21st of August, 2017. It will be visible in Oregon, Idaho, Wyoming, Nebraska, Kansas, Missouri, Illinois, Kentucky, Tennessee, Georgia, North Carolina, and South Carolina. The event will be the only total solar eclipse for Americans this decade.

FUTURE AMERICAN ECLIPSES

The next total eclipse to cross the continental United States is on April 8, 2024. It will travel from Texas to Maine. After that, the next American total eclipses will be in 2044 and 2045.

Viewing and Photographing the Eclipse

AT-HOME PINHOLE METHOD

Use the pinhole method to view the eclipse safely. It costs little but is the safest technique there is. Take a stiff piece of single-layer cardboard and punch a clean pinhole. Let the sun shine through the pinhole onto another piece of cardboard. That's it!

Never look at the sun through the pinhole. Your back should be toward the sun to protect your eyes. To brighten the image, simply move the back piece of cardboard closer to the pinhole. To see it larger, move the back cardboard farther away. Do not make the pinhole larger. It will only distort the crescent sun.

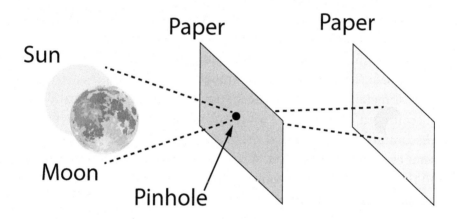

WELDING GOGGLES

Welding goggles that have a rating of fourteen or higher are another useful eclipse viewing tool. The goggles can be used to view the solar eclipse directly. Do not use the goggles to look through binoculars or telescopes, as the goggles could potentially shatter due to intense direct heat. Avoid long periods of gazing with the goggles. Look away every so often. Give your eyes a break.

SOLAR FILTERS FOR TELESCOPES

The ONLY safe way to view solar eclipses using telescopes or binoculars is to use solar filters. The filters are coated with metal to diminish the full intensity of the sun. Although the filters can be expensive, it is better to purchase a quality filter rather than an inexpensive one that could shatter or melt from the heat.

The filters attach to the front of the telescope for easy viewing. Remember to give your telescope cooling breaks. Rapid heating can damage your equipment with or without filters attached.

WATCH OUT FOR UNSAFE FILTERS

There are several myths surrounding solar filters for eclipse viewing. In order for filters to be safe, they must be specially designed for looking at a solar eclipse. The following are all unsafe for eclipse viewing and can lead to retinal damage: developed colored or chromogenic film, black-and-white negatives such as X-rays, CDs with aluminum, smoked glass, floppy disk covers, black-and-white film with no silver, sunglasses, or polarizing films.

Some online articles state that using developed black-and-white film is safe. Those articles fail to mention the film must have a layer of real metallic silver to protect your eyes. Using developed film is discouraged. You cannot ensure the quality of the film. Feeling no discomfort while looking at the partial solar eclipse does NOT mean your eyes are protected. Retinal damage can occur with zero pain due to the retinas having no pain receptors. Please be careful. Only use protective glasses certified for viewing the eclipse.

VIEWING WITH BINOCULARS

When viewing the eclipse with binoculars, it is important to use solar filters on both lenses until totality. Only then is it safe to remove the filter. As the sun becomes visible after totality, replace the filters for safe viewing. Protect your pupils. Remember to give your binoculars a cool-down break between viewings. They can overheat rapidly from being pointed directly at the sun even with filters attached.

PLANNING AHEAD

There are many things to keep in mind when viewing a total eclipse. It is important to plan ahead to get the most out of this extraordinary experience.

UNDERSTANDING SUN POSITION

All compass bearings in this book are true north. All compasses point to Earth's magnetic north. The difference between these two measurements is called magnetic declination. The magnetic declination for Missouri in August 2017 is:

0° 3' W ± 0° 21' (for Jefferson City)

Subtract the declination from the azimuth bearing as given in the text, and set your compass to that direction.

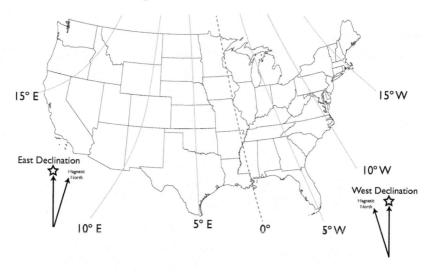

If you purchase a compass with a built-in declination adjustment, you can change the setting once and eliminate the calculations. The Suunto M-3G compass has this correction. A compass with a sighting mirror or wire will help you make a more accurate azimuth sighting.

The Suunto M-3G also has an inclinometer. This allows you to measure the elevation of any object above the horizon. Use this to figure out how high the sun will be above your position.

You can also use a smartphone inclinometer and compass for this purpose. Make sure to calibrate your smartphone's compass before every use, otherwise it might indicate the wrong bearing. Set the smartphone compass for true north to match the book. Understand the compass prior to August 21. There will be little time to guess or search on Google. Smartphone and GPS compasses are "sticky." Their compasses don't swing as freely as a magnetic compass does.

The author has used his magnetic compass for azimuth measurements and a smartphone to measure elevation. Combining these two tools will allow you to make the best sightings possible.

Outdoor sporting goods stores in most towns and cities carry compasses. We recommend purchasing a good compass in your hometown. Take the time to learn how to use it before the day of the eclipse. You do not want to struggle with orienteering basics under pressure.

SUN AZIMUTH

Azimuth is the compass angle along the horizon, with 0° corresponding to north, and increasing in a clockwise direction. 90° is east, 180° is south, and 270° is west.

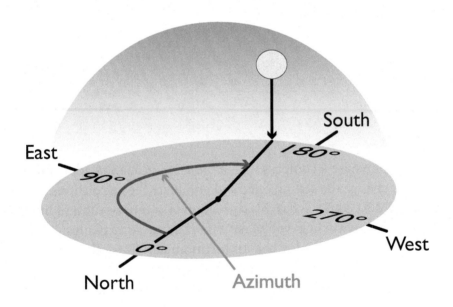

SUN ELEVATION

Altitude is the sun's angle up from the horizon. A 0° altitude means exactly on the horizon and 90° means "straight up."

Using the sun azimuth and elevation data, you can predict the position of the sun at any given time. Positions given in this book coincide with the time of eclipse totality unless otherwise noted.

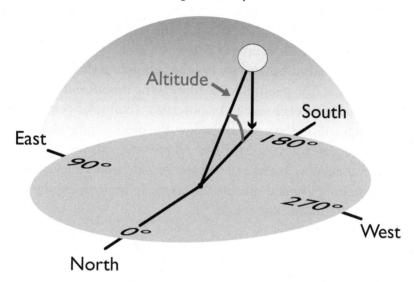

ECLIPSE DATA FOR SELECT MISSOURI LOCATIONS

LOCATION	TOTALITY START (CDT)
BOONVILLE	1:11:26PM
COLUMBIA	1:12:21PM
JEFFERSON CITY	1:13:06PM
KANSAS CITY	1:08:30PM
ST. JOSEPH	1:06:27PM
ST. LOUIS	NONE

ECLIPSE PHOTOGRAPHY

Photographing an eclipse is an exciting challenge, as the moon's shadow moves near 2,000MPH. There is an element of danger and the pressure of time. Looking at the unfiltered sun through a camera can permanently damage your vision and your camera. If you are unsure, just enjoy the eclipse with specially designed glasses. Keep a solar filter on your lens during the eclipse and remove for the duration of totality!

Partial Vs. Total Solar Eclipse

To successfully and safely photograph a partial and total eclipse, it is important to understand the difference between the two. A solar eclipse occurs when the moon is positioned between the sun and Earth. The region where the shadow of the moon falls upon Earth's surface is where a solar eclipse is visible.

The moon's shadow has two parts—the penumbral shadow and the umbral shadow. The penumbral shadow is the moon's outer shadow where partial solar eclipses can be observed. Total solar eclipses can only be seen within the umbral shadow, the moon's inner shadow.

You cannot say you've seen a total eclipse when all you saw was a partial solar eclipse. It is like saying you've watched a concert, but in reality, you only listened outside the arena. In both cases, you have missed the drama and the action.

Photographing A Partial And Total Solar Eclipse

Aside from the region where the outer shadow of the moon is cast, a partial solar eclipse is also visible before a total solar eclipse within

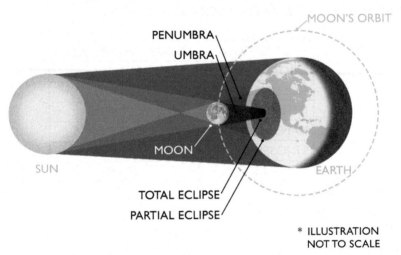

PENUMBRA
UMBRA
MOON'S ORBIT
MOON
SUN
EARTH
TOTAL ECLIPSE
PARTIAL ECLIPSE

* ILLUSTRATION
NOT TO SCALE

the inner shadow region. In both cases, it is imperative to use a solar filter on the lens for both photography and safety reasons. This is the only difference between taking a partial eclipse and a total eclipse photograph of the sun.

To photograph a total solar eclipse, you must be within the Path of Totality, the surface of the Earth within the moon's umbral shadow.

The Challenge

A total solar eclipse only lasts for a couple of minutes. It is brief, but the scenario it brings is unforgettable. Seeing the radiant sun slowly being covered by darkness gives the spectator a high level of anticipation and indescribable excitement. Once the moon completely covers the sun's radiance, the corona is finally visible. In the darkness, the sun's corona shines, capturing the crowd's full attention. Watching this phenomenon is a breathtaking experience.

Amidst all the noise, cheering, and excitement, you have less than 150 seconds to take a perfect photograph. The key to this is planning. You need to plan, practice, and perfect what you will do when the big moment arrives because there is no replay. The pressure is enormous. You only have two minutes to capture the totality and the sun's corona using different exposures.

Plan, Practice, Perfect

It is important to practice photographing before the actual phenomenon arrives. Test your chosen imaging setup for flaws. Rehearse over and over until your body remembers what you will do from the moment you arrive at your chosen spot to the moment you pack up and leave the area.

You will discover potential problems regarding vibrations and focus that you can address immediately. This minimizes the variables that might affect your photographs at the most critical moment.

It's common for experienced eclipse chasers to lose track of what they plan to do. Write down what you expect to do. Practice it time and again. Play annoying, distracting music while you practice. Try photographing in the worst weather possible. Do anything you can to practice under pressure. Eclipse day is not the time to practice.

Once the sun is completely covered, don't just take photographs. Capture the experience and the image of the total solar eclipse in your mind as well. Set up cameras around you to record not just the total solar eclipse but also the excitement and reaction of the crowd.

ECLIPSE PHOTOGRAPHY GEAR

What do you need to photograph the total eclipse? There are only a few pieces of equipment that you'll need. Preparing to photograph an eclipse successfully takes time. Not only do you have to be skilled and have the right gear, you have to be in the correct place.

BASIC ECLIPSE PHOTOGRAPHY EQUIPMENT

- Solar viewing glasses
- Lens solar filter
- Minimum 300mm lens
- Stable tripod that can be tilted to 60° vertical
- High-resolution DSLR
- Spare batteries for everything
- Secondary camera to photograph people, the horizon, etc.
- Remote cable or wireless release

ADDITIONAL ITEMS

- Video camera
- Video camera tripod
- Quality pair of binoculars
- Solar filters for each binocular lens
- Photo editing software

EQUIPMENT TO PREPARE BEFORE THE BIG DAY

A. Solar viewing glasses

You need a pair of solar viewing glasses as the eclipse approaches.

B. Solar Filter

Partial and total eclipse photography is different from normal photography. Even if only 1% of the sun's surface is visible, it is still approximately 10,000 times brighter than the moon. Before totality, use a solar filter on your lens. Do not look at the sun with your eyes. It can cause irreparable damage to your retinas.

DO NOT leave your camera pointed at the sun without a solar filter attached. The sun will melt the inside of your camera. Think of a magnifying glass used to torch ants and multiply that by one hundred.

C. Lens

To capture the corona's majesty, you need to use a telescope or a telephoto lens. The best focal length, which will give you a large image of the sun's disk, is 400mm and above. You don't want to waste all your efforts by bringing home a small dot where the black disk and majestic corona are supposed to be.

D. Tripod

Bring a stable enough tripod to support your camera properly to avoid unsteady shots and repeated adjustments. Either will ruin your photos. It also needs to be portable in case you need to change locations for a better shot.

E. Camera

You need to remember to set your camera to its highest resolution to capture all the details. Set your camera to:

- 14-bit RAW is ideal, otherwise
- JPG, Fine compression, Maximum resolution

Bracket your exposures. Shoot at various shutter speeds to capture different brightnesses in the corona. Note that stopping your lens all the way down may not result in the sharpest images.

Choose the lowest possible ISO for the best quality while maintaining a high shutter speed to prevent blurred shots. Set your camera to manual. Do not use AUTO ISO. Your camera will be fooled. The night before, test the focus position of your lens using a bright star or the moon.

Constantly double-check your focus. Be paranoid about this. You can deal with a grainy picture. No amount of Photoshop will fix a blurry, out-of-focus picture.

F. Batteries

Remember to bring fresh batteries! Make sure that you have enough power to capture the most important moments. Swap in fresh batteries thirty minutes before totality.

G. Remote release

Use a wired or wireless remote release to fire the camera's shutter. This will reduce the amount of camera vibration.

H. Video Camera

Run a video camera of yourself. Capture all the things you say and do during the totality. You'll be amazed at your reaction.

I. Photo editing software

You will need quality photo editing software to process your eclipse images. Adobe Lightroom and Photoshop are excellent programs to extract the most out of your images. Become well versed in how to use them at least a month before the eclipse.

J. Smartphone applications

The following smartphone applications will aid in your photography planning: Wunderground, Skyview, Photographer's Ephemeris, Sunrise and Sunset Calculator, SunCalc, and Sun Surveyor among others.

CAMERA PHONES

Smartphone cameras are useful for many things but not eclipse photography. An iPhone 6 camera has a 63° horizontal field of view and is 3264 pixels across. If you attempt to photograph the eclipse, the sun will be a measly 30-40 pixels wide depending on the phone. Digital pinch zoom won't help here. If you want *National Geographic* images, you'll need a serious camera and lens, far beyond any smartphone.

Consider instead using a smartphone to run a time-lapse of the entire event. The sun will be minuscule when shot on a smartphone. Think of something else exciting and interesting do to with it. Purchase a Gorilla Pod, inexpensive tripod, or selfie stick and mount the smartphone somewhere unique.

Also, partial and total eclipse light is strange and ethereal. Consider using that light to take unique pictures of things and people. It's rare and you may have something no one else does.

FOCAL LENGTH & THE SIZE OF SUN

The size of the sun in a photo depends on the lens focal length. A 300mm lens is the recommended minimum on a full-frame (FF) DSLR. Lenses up to this size are relatively inexpensive. For more magnification, use an APS-C (crop) size sensor. Cameras with these sensors provide an advantage by capturing a larger sun.

For the same focal length, an APS-C sensor will provide a greater apparent magnification of any object. As a consequence, a shorter, less expensive lens can be used to capture the same size sun.

The below figure shows the size of the sun on a camera sensor at various focal lengths. As can be seen with the 200mm lens, the sun is quite small. On a full-frame camera at 200mm, the sun will be 371 pixels wide on a Nikon D810, a 36-megapixel body. A lower resolution FF camera will result in an even smaller sun.

Printing a 24-inch image shot on a Nikon D810 with a 200mm lens at a standard 300 pixels per inch results in a small sun. On this size paper, the sun will be a miserly 1.25 inches wide!

Photographing the eclipse with a lens shorter than 300mm will leave you with little to work with. Using a 400mm lens and printing a 24-inch print will result in a 2.5-inch-wide sun. For as massive as the sun is, it is a challenge to take a photograph with the sun of any meaningful size.

FOCAL LENGTH	FOV FULL FRAME	FF VERT. ANGLE	% OF FF	SUN PIXEL SIZE
14	104° X 81°	81°	0.7%	32.1
20	84° X 62°	62°	0.9%	41.9
28	65° X 46°	46°	1.2%	56.5
35	54° X 38°	38°	1.4%	68.5
50	40° X 27°	27°	2.0%	96.4
105	19° X 13°	13°	4.1%	200.2
200	10° X 7°	7°	7.6%	371.9
400	5° X 3.4°	3.4°	15.6%	765.6
500	4° X 2.7°	2.7°	19.6%	964.2
1000	2° X 1.3°	1.3°	40.8%	2002.5
1500	1.4° X 0.9°	0.9°	58.9%	2892.6
2000	1° X 0.68°	0.68°	77.9%	3828.4

Chart 1: Full-frame camera field of view. The 3rd column is the vertical field of view in degrees. Column 4 is the percentage of the total sensor height that the sun covers. Column 5 is how many pixels wide the sun will be on a 36MP Nikon D810. (Values are estimates)

FOCAL LENGTH	FOV CROP	CROP VERT DEG	% OF CROP	SUN PIXEL SIZE
14	80° X 58°	58°	0.9%	33.9
20	61° X 43°	43°	1.2%	45.8
28	45° X 31°	31°	1.7%	63.5
35	37° X 25°	25°	2.1%	78.7
50	26° X 18°	18°	2.9%	109.3
105	13° X 8°	8°	6.6%	245.9
200	6.7° X 4.5°	4.5°	11.8%	437.2
400	3.4° X 2°	2°	26.5%	983.7
500	2.7° X 1.8	1.8°	29.4%	1093.0
1000	1.3° X 0.9°	0.9°	58.9%	2186.0
1500	0.9° X 0.6°	0.6°	88.3%	3278.9
2000	0.6° X 0.45°	0.5°	117.8%	4371.9

Chart 2: APS-C Crop sensor camera field of view. The 3rd column is the vertical field of view in degrees. Column 4 is the percentage of the total sensor height that the sun covers. Column 5 is how many pixels wide the sun will be on a 12mp Nikon D300s. (Values are estimates)

The big challenge is the cost of the lens. Lenses longer than 300mm are expensive. They also require heavier tripods and specialized tripod heads. The 70-300mm lenses from Nikon, Canon, Tamron, and others are relatively affordable options. It is worth spending time at a local camera shop to try different lenses. Long focal-length lenses are a significant investment, especially for a single event.

To achieve a large eclipse image, you will need a long focal-length lens, ideally at least 400mm. A standard 70-300mm lens set to 300mm will show a small sun. At 500mm, the sun image becomes larger and covers more of the sensor area. The corona will take up a significant portion of the frame. By 1000mm, the corona will exceed the capture area on a full-frame sensor. See the picture on page thirty-seven for sun size simulations for different focal lengths.

SUGGESTED EXPOSURES

To photograph the partial eclipse, the camera must have a solar filter attached. If not, the intense light from the sun may damage (fry) the inside of your camera. This has happened to the author. The exposure depends on the density (darkness) of the solar filter used.

As a starting point, set the camera to ISO 100, f/8, and with the solar filter on, try an exposure of 1/4000. Make adjustments based on the histogram and highlight warning.

Turn on the highlight warning in your camera. This feature is commonly called "blinkies." This warning will help you detect if the image is overexposed or not.

Once the Baily's Beads, prominences, and corona become visible, there will only be 2.25 minutes to take bracketed shots. It will take at least eleven shots to capture the various areas of the sun's corona. The brightness varies considerably. No commercially available camera can capture the incredible dynamic range of the different portions of the delicate corona. This requires taking multiple photographs and digitally combining them afterward.

During totality, try these exposure times at ISO 100 and f/8:

1/4000, 1/2000, 1/1000, 1/250, 1/60, 1/30, 1/15, 1/4, 1/2, 1 sec, and 4 sec.

PHOTOGRAPHY TIME

Set the camera to full-stop adjustments. It will reduce the time spent fiddling. As an example, the author tried the above shot sequence, adjusting the shutter speed as fast as possible.

It took thirty-three seconds to shoot the above 11 shots using 1/3-stop increments. This was without adjusting composition, focus, or anything else but the shutter speed. When the camera was set to full stop increments, it only took twenty-two seconds to step through the same shutter speed sequence.

Assuming the totality lasts less than two minutes, only four shot sequences could be made using 1/3-stop increments. Yet six shot sequences could be made when the camera was set to full stop steps. Zero time was spent looking at the back LCD to analyze highlights and the histogram.

Now add in the bare minimum time to check the highlight warning. It took sixty-three seconds to shoot and check each image using full stops. And that was without changing the composition to allow for sun movement, bumping the tripod, etc. Looking at the LCD ("chimping") consumed **half** of the totality time.

This test was done in the comfort of home under no pressure. In real world conditions, it may be possible to successfully shoot only one sequence. If you plan to capture the entire dynamic range of the totality, you must practice the sequence until you have it down cold. If you normally fumble with your camera, do not underestimate the difficulty, frustration, and stress of total eclipse photography.

Most importantly, trying to shoot this sequence allowed for zero time to simply look at the totality to enjoy the spectacle.

AVOID LAST MINUTE PURCHASES

You should purchase whatever you think you'll need to photograph the eclipse today. This event will be nothing short of massive. Remember the hot toy of the year? Multiply that frenzy by a thousand. Everyone will want to try to capture their own photo.

Do not wait until the last few weeks before the eclipse to purchase cameras, lenses, filters, tripods, viewing glasses, and associated material. Consider that the totality of the eclipse will streak from coast to coast. Everyone who wants to photograph the eclipse will order at the same time. If you wait until August to buy what you need, it's conceivable that every piece of camera equipment capable of creating a total eclipse photo will be sold out in the United States. Whether this happens or not, do not wait until midsummer to make your purchases. It may be too late.

PRACTICE

You will need to practice with your equipment. Things may go wrong that you don't anticipate. If you've never photographed a partial or total eclipse, taking quality shots is more difficult than you think. Practice shooting the sequence with a midday sun. This will tell you if you have your exposures and timing correct. Figure out what you need well in advance.

Practice photographing the full moon and stars at night. Capture the moon in full daylight. There will be six moon cycles to practice with. Astrophotography is challenging and requires practice.

The May 20, 2012, eclipse as seen in San Diego, CA, shot with a Nikon D300s (crop sensor) with an 80-400mm lens set to ~350mm. The sun is 560 pixels wide on the 4288x2848 image.

This image is shown straight out of the camera without modification. Even with a high-quality camera and lens, photographing an eclipse is challenging. Note the haze and reflection from the overexposed sun.

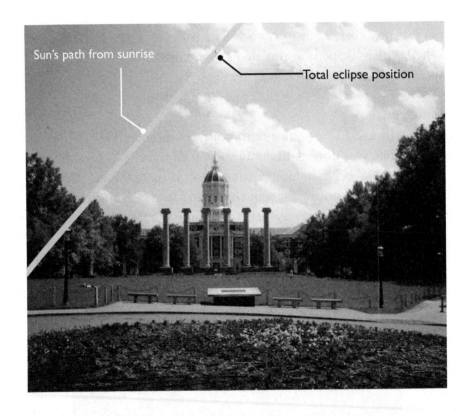

The sun will follow this path on the morning of the eclipse on August 21, 2017. Image of Francis Quandrangle, University of Missouri, Columbia, Missouri.

Hopefully there will not be many clouds on the day of the eclipse. *Note that this image is a simulation and approximation of where the total eclipse will appear. Refer to the eclipse position data for a more accurate location.*

◉ the symbol for the sun and first appeared in Europe during the Renaissance.

☾ is the ancient symbol for the moon.

Viewing Locations Around Missouri

Hundreds of thousands of people will travel to and around Missouri to view the total eclipse. The skies are partly clear, and there is a vast amount of space to view the total eclipse from.

If the weather is questionable, seek out a new location as soon as possible. If you wait until the hour before the eclipse, you may find yourself stuck in traffic, as others will be looking for clear skies. Be safe on the roadways, as drivers may be distracted.

This section contains popular, alternative, and little-known locations to watch the eclipse. As long as there are no clouds or smoke from fires, the partial eclipse will be viewable from anywhere in Missouri.

Suggested Total Eclipse View Points

Towns and Cities

- Boonville
- Chamois
- Columbia
- Excelsior Springs
- Festus
- Fulton
- Jefferson City
- Kansas City
- Marshall
- Plattsburg
- Sedalia
- St. Clair
- St. Joseph
- St. Louis
- Ste. Genevieve
- Sullivan

Jefferson City

Missouri Total Eclipse Path

Unique Locations

- Arrow Rock SHS
- Battle of Lexington SHS
- Mark Twain National Forest
- Missouri River
- Washington SP

Boonville

Elevation:	667 feet
Population:	8,370 (2013)
Main road/hwy:	I-70

Overview

As with other cities along the centerline of the totality, many hotels are already booked. As more people find out about the eclipse, hotel rooms will become difficult to find. Boonville is no exception, according to their Chamber of Commerce eclipse website. Boonville plans to have a block party on Sunday at the Depot District with music, food, and games. The State Fair Community College with have a BBQ on the day of the eclipse.

Getting There

Drive east on I-70 from Kansas City for 103 miles to reach the city of Boonville.

Totality Duration

2 minutes 39 seconds

Notes

Visit the Boonville Chamber of Commerce website for updates and more information:

boonvillemochamber.com/missouri-solar-eclipse-2017

Event	Time (CDT)	Altitude	Azimuth
Sunrise	6:30:00AM	0°	73°
Eclipse Start	11:44:52AM	56°	137°
Totality Start	1:11:26PM	62°	178°
Totality End	1:14:06PM	62°	180°
Eclipse End	2:39:27PM	56°	220°
Sunset	7:57:00PM	0°	285°

CHAMOIS

Elevation:	538 feet
Population:	396 (2017)
Main road/hwy:	Hwy 100

Chamois

OVERVIEW

Chamois is a city in Osage County. The city was platted in 1856, and named after the chamois, a European animal. The town's post office has been in continuous operation since before the Civil War in 1856.

Chamois is one of several Missouri communities nearly on the centerline of the total eclipse on August 21.

GETTING THERE

Drive east on US 50 from Jefferson City for nineteen miles. Exit onto Highway 100 and drive north eighteen miles to reach Chamois.

TOTALITY DURATION

2 minutes 38 seconds

NOTES

There is a photo essay by Gen Baugher featuring Chamois. It can be found at

www.myphotoessays.com/GenBaugher/Chamois.html.

Event	Time (CDT)	Altitude	Azimuth
Sunrise	6:26:00AM	0°	73°
Eclipse Start	11:46:52AM	57°	139°
Totality Start	1:13:46PM	63°	182°
Totality End	1:16:25PM	63°	183°
Eclipse End	2:41:37PM	56°	223°
Sunset	7:52:00PM	0°	285°

COLUMBIA

Elevation:	758 feet
Population:	115,276 (2013)
Main road/hwy:	I-70

OVERVIEW

Columbia is one of the largest cities in Missouri along the path of totality. The city's main viewing event will be at Cosmo Park. They plan to have food, activities, and music celebrating the day of the eclipse. Like other locations in Missouri, the city lies in the area of the greatest duration of the eclipse.

GETTING THERE

Columbia is a major city in the center of Missouri. It can be reached from either St. Louis or Kansas City along the I-70.

TOTALITY DURATION

2 minutes 35 seconds

NOTES

Visit the Columbia website for more information about updates and events in the city:

www.visitcolumbiamo.com/events/total-solar-eclipse

The city also has a Facebook page for the event:

www.facebook.com/columbiamissourieclipse2017

Event	Time (CDT)	Altitude	Azimuth
Sunrise	6:28:00AM	0°	73°
Eclipse Start	11:45:41AM	56°	138°
Totality Start	1:12:21PM	62°	180°
Totality End	1:14:58PM	62°	181°
Eclipse End	2:40:13PM	56°	221°
Sunset	7:55:00PM	0°	285°

Excelsior Springs

Elevation:	784 feet
Population:	11,084 (2010)
Main road/hwy:	US 69

Overview

Excelsior Springs is a vibrant community of eleven thousand people. The city has a unique history. Its natural beauty makes it the ideal place to view the 2017 solar eclipse. Its numerous mineral wells in the late 1800s attracted people to come and experience "America's Haven of Health." The Chamber of Commerce and the Downtown Excelsior Partnership (DEP) are working with the city's museum to create an event that will showcase the city's historic places and festivals.

For details visit visitexcelsior.com/SolarFest.

Getting There

Drive north from Kansas City on Interstate 35 and exit on US 69.

Totality Duration

2 minutes 33 seconds

Notes

The author's grandfather Roy "Akira" Takeda was placed in a Japanese internment camp during World War II but was able to leave when he found work as a dishwasher at the Elms Hotel in Excelsior Springs. He climbed to management and enjoyed a successful fifty years in the hospitality industry.

Event	Time (CDT)	Altitude	Azimuth
Sunrise	6:35:00AM	0°	73
Eclipse Start	11:41:55AM	55°	135°
Totality Start	1:08:06PM	62°	173°
Totality End	1:10:39PM	62°	175°
Eclipse End	2:36:15PM	57°	216°
Sunset	8:03:00PM	0°	286°

FESTUS

Elevation:	435 feet
Population:	11,602 (2010)
Main road/hwy:	I-55

OVERVIEW

The eclipse event for Festus city will take place at West City Park, Eclipse Central, among other viewing points. West City Park is on Sunshine Dr. (Old Highway A for those utilizing a guide or route map) West City Park is an area with over one hundred acres of open space. It offers a great view of the sky. Festus plans to have viewing points from the fairgrounds, stage field, and northern ball fields. Parking space will be available at the ball field complex. Informational presentations and countdown to the eclipse will happen at West City Park. Activities planning for the whole weekend are underway. Eclipse events are supported by the Festus Tourism Commission with support from the Festus Public Library.

GETTING THERE

Drive south from St. Louis on Interstate 55 for thirty-four miles.

TOTALITY DURATION

2 minutes 35 seconds

NOTES

Visit the Festus eclipse website for updated information: www.festuseclipse.com

Event	Time (CDT)	Altitude	Azimuth
Sunrise	6:21:00AM	0°	74°
Eclipse Start	11:49:60AM	58°	142°
Totality Start	1:17:06PM	63°	186°
Totality End	1:19:41PM	63°	188°
Eclipse End	2:44:40PM	55°	226°
Sunset	7:46:00PM	0°	285°

FULTON

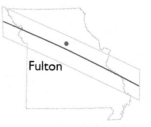

Elevation:	764 feet
Population:	12,770 (2010)
Main road/hwy:	US 54

Fulton

OVERVIEW

Fulton is planning a series of events leading up to the total eclipse sweeping across the state of Missouri. Dr. Angela Speck of the University of Missouri visited Fulton to meet with the local committee that is planning activities for the 2017 solar eclipse. The Winston Churchill Memorial and Library (The National Churchill Museum) is located on the Westminster College campus in Fulton. The National Churchill Museum is marked by three related elements: the museum, St. Mary Aldermanbury church, and the Breakthrough sculpture. A series of events is set to take place from Friday, August 18 to Monday, August 21. For more details, visit their Facebook event page at www.facebook.com/events/1847690805498617/.

GETTING THERE

Drive north from Jefferson City on US 54 for twenty-four miles to reach Fulton.

TOTALITY DURATION

2 minutes 32 seconds

NOTES

Visit the city's eclipse webpage at www.visitfulton.com/eclipse.html.

Event	Time (CDT)	Altitude	Azimuth
Sunrise	6:27:00AM	0°	73°
Eclipse Start	11:46:26AM	57°	139°
Totality Start	1:13:15PM	63°	181°
Totality End	1:15:47PM	63°	182°
Eclipse End	2:41:00PM	56°	222°
Sunset	7:53:00PM	0°	285°

Jefferson City

Elevation: 630 feet
Population: 43,330 (2013)
Main road/hwy: US 54/63

Jefferson City

Overview

Jefferson City is one of several state capitals to be under the path of totality on August 21. It is also one of the larger cities to enjoy the greatest duration of the eclipse in the United States. The city plans to have a downtown street party, educational events at several locations, and even a Pink Floyd tribute band playing on the weekend of the eclipse. Riverside Park will host events all weekend and on the day of the eclipse.

Getting There

Drive west on I-70 from St. Louis and exit on US 54. Travel south on US 54 until you reach Jefferson City.

Totality Duration

2 minutes 29 seconds

Notes

Visit Jefferson City's official eclipse website for more information about events:

www.capitaleclipse.org

Event	Time (CDT)	Altitude	Azimuth
Sunrise	6:28:00PM	0°	73°
Eclipse Start	11:46:06AM	57°	138°
Totality Start	1:13:06PM	63°	180°
Totality End	1:15:34PM	63°	182°
Eclipse End	2:41:01PM	56°	222°
Sunset	7:54:00PM	0°	285°

KANSAS CITY

Elevation:	909 feet
Population:	475,378 (2015)
Main road/hwy:	Multiple

Kansas City

OVERVIEW

Kansas City is the largest city in the state of Missouri. With a metropolitan population of 2.1 million, greater Kansas City covers a fifteen-county area. Although it is one of the largest cities in the path of totality, only the northeast portion of the metro area will see the total eclipse. The city is a good starting point to reach other eclipse locations in Missouri, Kansas, and Nebraska. The cities of Independence, North Kansas City, Parkville, Liberty, Platte Woods, Ferrelview, and Smithville will experience a longer totality than the areas in the core of Kansas City.

GETTING THERE

Kansas City is a major metropolitan area in west Missouri. Drive or fly there from anywhere in the region.

TOTALITY DURATION

27 seconds

NOTES

Travel northeast along I-35 to find a longer totality.

Event	Time (CDT)	Altitude	Azimuth
Sunrise	6:37:00AM	0°	73°
Eclipse Start	11:41:19AM	54°	134°
Totality Start	1:08:30PM	62°	173°
Totality End	1:09:14PM	62°	173°
Eclipse End	2:35:59PM	58°	216°
Sunset	8:04:00PM	0°	285°

MARSHALL

Elevation:	764 feet
Population:	13,065 (2010)
Main road/hwy:	US 65

Marshall

OVERVIEW

Marshall has multiple events planned for the 2017 total eclipse. A cosmic concert is set to happen on the weekend. Activities are set to take place in two prime public viewing points: Indian Foothills Park, 1538 E. Vest Street, and Salt Fork YMCA, 740 E. Yerby Street. Both are excellent viewing points. Complimentary parking is also available in the heart of Marshall. Activities to take place include Shopping Under the Sun, a 5K walk/run, Night & Day Concert under the Stars, and camping. The Marshall Solar Eclipse Committee will offer solar eclipse education sessions in the spring to build comprehension about the eclipse. The sessions will focus on how and why and eclipse happens, eclipse safety, telescope use, and much more!

GETTING THERE

Drive west from Columbia on I-70 and turn north on US 65 to Marshall.

TOTALITY DURATION

2 minutes 39 seconds

NOTES

Visit the city's eclipse website: www.marshallmosolareclipse.com

Event	Time (CDT)	Altitude	Azimuth
Sunrise	6:31:00AM	0°	73°
Eclipse Start	11:43:56AM	56°	137°
Totality Start	1:10:22PM	62°	177°
Totality End	1:13:01PM	62°	178°
Eclipse End	2:38:24PM	57°	219°
Sunset	7:59:00PM	0°	285°

PLATTSBURG

		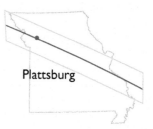
Elevation:	948 feet	
Population:	2,319 (2010)	
Main road/hwy:	Hwy 116	

OVERVIEW

Located only a few miles south of the eclipse centerline, Plattsburg should be an excellent viewing location for the total eclipse with over a 2.5-minute totality. Located north of Smithville Lake, Plattsburg is a good starting point to find your ideal eclipse viewing point. The entire area is laced with creeks to cool off in. Perkins Park, which is southeast of town, will also be another fine area to enjoy the total eclipse from.

GETTING THERE

Drive north from Kansas City on US 169 for thirty-three miles. Turn east on Highway 116 and drive 6.6 miles to reach Plattsburg.

TOTALITY DURATION

2 minutes 38 seconds

NOTES

For any upcoming events leading to the eclipse, check the town's official website:

www.plattsburgmo.com

Event	Time (CDT)	Altitude	Azimuth
Sunrise	6:36:00AM	0°	73°
Eclipse Start	11:41:26AM	54°	134°
Totality Start	1:07:24PM	62°	172°
Totality End	1:10:02PM	62°	174°
Eclipse End	2:35:32PM	57°	215°
Sunset	8:04:00PM	0°	286°

Sedalia

Elevation:	909 feet
Population:	21,387 (2010)
Main road/hwy:	US 50

Overview

Sedalia is considered a prime viewing location in Missouri. The Missouri State Fair Foundation has planned festivities that will include a horseback trail ride, Cow Patty Bingo, and other events. Sedalia is working along with organizations in Clay County and the Old Trails Region of northwest Missouri to plan eclipse-related events.

The Sedalia Visitors Bureau site is www.cityofsedalia.com.

They also have a Facebook page with the latest updates at www.facebook.com/events/1882767598649146.

Getting There

Drive west from Jefferson City sixty miles on US 50 to reach Sedalia.

Totality Duration

1 minutes 35 seconds

Notes

Visit the town's website for updates during the summer: ci.sedalia.mo.us

Event	Time (CDT)	Altitude	Azimuth
Sunrise	6:32:00AM	0°	73°
Eclipse Start	11:43:59AM	56°	136°
Totality Start	1:11:13PM	63°	177°
Totality End	1:12:51PM	63°	178°
Eclipse End	2:38:59PM	57°	220°
Sunset	7:58:00PM	0°	285°

St. Clair

Elevation:	761 feet
Population:	4,472 (2010)
Main road/hwy:	I-44

St. Clair

Overview

St. Clair is set to serve viewers with comforts as well as the greatest views on the continent. St. Clair has planned activities from Friday, August 18 to Monday, August 21. St. Clair city parks will offer movies in the park. A bluegrass festival will be held at Orchard Park on Saturday. Visit their website: www.jumpstartstclair.com/bluegrass-festival. The Chamber of Commerce will sponsor a car show on Route 66. The Airstrip, Orchard Park, and the Plaza Shops on Route 66 will offer prime viewing points. The VFW Post is the best spot to view the eclipse on Route 66. Lost Hill Lake Events in St. Clair, Missouri has large fields available for viewing the eclipse. Lost Hill Lake is five miles from the totality line. https://losthilllakeevents.com/eclipse/

Getting There

Drive southwest from St. Louis on Interstate 44 for fifty-two miles to reach St. Clair.

Totality Duration

2 minutes 39 seconds

Notes

Visit St. Clair's website for updates at www.stclaireclipse.com.

Event	Time (CDT)	Altitude	Azimuth
Sunrise	6:24:00AM	0°	74°
Eclipse Start	11:48:32AM	58°	141°
Totality Start	1:15:42PM	63°	184°
Totality End	1:18:22PM	63°	186°
Eclipse End	2:43:27PM	55°	225°
Sunset	7:49:00PM	0°	285°

St. Joseph

Elevation:	948 feet
Population:	77,147 (2013)
Main road/hwy:	I-29

Overview

St. Joseph is planning a major event at the St. Joseph airport. The senior editor of *Astronomy* magazine will be hosting this gathering, which is expected to draw as many as one hundred thousand people. The city lies directly on the centerline of the eclipse. There are several websites with specific information and updates for St. Joseph. The St. Joseph Convention and Visitors Bureau site is: www.StJoMoEclipse.com. They also have a Facebook page with the latest updates:

www.facebook.com/Eclipse2017inStJoMo

Getting There

St. Joseph is a large city in Missouri. Drive north from Kansas City on I-29 for fifty-five miles to reach St. Joseph. Alternatively, fly into the city several days prior to the eclipse.

Totality Duration

2 minutes 38 seconds

Notes

Visit the eclipse event website for updates and more information: stjosepheclipse.com

Event	Time (CDT)	Altitude	Azimuth
Sunrise	6:37:00AM	0°	73°
Eclipse Start	11:40:42AM	54°	134°
Totality Start	1:06:27PM	61°	171°
Totality End	1:09:08PM	61°	173°
Eclipse End	2:34:36PM	57°	214°
Sunset	8:06:00PM	0°	286°

St. Louis

Elevation:	465 feet	
Population:	318,416 (2013)	
Main road/hwy:	Multiple	

St. Louis

Overview

As with Kansas City, St. Louis is one of the largest cities in the path of totality across the United States. However, it also on the edge of the total eclipse shadow. The southwestern metro area of St. Louis will experience the total eclipse while the central and northern sections of the city will only see a partial eclipse. The metro area of St. Louis has a population of 2.8 million, making it one of the largest cities in the region. The greater city area also covers part of Illinois as well.

Getting There

St. Louis is a major city in Missouri. Drive or fly there from any city in the region.

Totality Duration

27 seconds or less

Notes

The cities of Chesterfield, Arnold, Pacific and others to the southwest of the city will experience the totality. Times listed below are for central St. Louis.

Event	Time (CDT)	Altitude	Azimuth
Sunrise	6:20:00AM	0°	73°
Eclipse Start	11:49:57AM	58°	143°
Totality Start	NONE	N/A	N/A
Totality End	NONE	N/A	N/A
Eclipse End	2:44:20PM	55°	226°
Sunset	7:46:00PM	0°	285°

Ste. Genevieve

Elevation:	560 feet
Population:	4,410 (2010)
Main road/hwy:	I-55

Ste. Genevieve

Overview

Ste. Genevieve is another excellent eclipse city. It has a perfect viewing area with few trees or buildings to obscure the views. Hawn State Park and the Hawn Association will host "Whispering Eclipse." The city has multiple activities planned for your weekend enjoyment. The St. Louis-based folk duo Riverside Wanderers will perform at the campground amphitheater on August 19. An informative session is set to take place on August 20. On Monday, a tent will be set up at the park entrance for shade and viewing of the eclipse.

To learn more about the eclipse and activities, visit mostateparks.com/event/66206/whispering-eclipse

Getting There

Drive south from St. Louis on Interstate 55. Take exit 154 and drive eight miles east to reach Ste. Genevieve.

Totality Duration

2 minutes 40 seconds

Notes

Visit the city's Facebook page for updates at www.facebook.com/stegen2017eclipse/

Event	Time (CDT)	Altitude	Azimuth
Sunrise	6:20:00AM	0°	74°
Eclipse Start	11:50:33AM	59°	143°
Totality Start	1:18:01PM	63°	188°
Totality End	1:20:41PM	63°	189°
Eclipse End	2:45:37PM	55°	227°
Sunset	7:45:00PM	0°	285°

SULLIVAN

Elevation:	930 feet
Population:	7,081 (2010)
Main road/hwy:	I-44

OVERVIEW

The city of Sullivan straddles the border of Franklin County and Crawford County in Missouri. This city is a good location to escape the crowds and heat of St. Louis during the weekend of the eclipse. You can visit Meramec State Park or City Lake Park to view the eclipse in natural surroundings. You'll be able to see the birds and insect think it's night, so they'll quiet down and sleep in these parks.

GETTING THERE

Drive southwest on Interstate 44 for sixty-seven miles to reach Sullivan.

TOTALITY DURATION

2 minutes 32 seconds

NOTES

Visit Sullivan's website for more eclipse information:
www.sullivan.mo.us/Eclipse/

Event	Time (CDT)	Altitude	Azimuth
Sunrise	6:24:00AM	0°	74°
Eclipse Start	11:48:15AM	58°	140°
Totality Start	1:15:35PM	63°	184°
Totality End	1:18:06PM	63°	185°
Eclipse End	2:43:24PM	56°	225°
Sunset	7:49:00PM	0°	285°

Arrow Rock SHS

Elevation:	709 feet	
Population:	56 (2013)	
Main road/hwy:	Hwy 41	

Arrow Rock SHS

Overview

The Arrow Rock community is formulating an eclipse-watching party. There will be a number of events in the village this weekend to anticipate and celebrate the total solar eclipse. On August 17, enjoy a night walk at nine p.m. Learn more and make reservations at friendsofarrowrock.org/nightwalks/.

On Sunday, there will be a stargazing event near the picnic shelter in the Arrow Rock State Historic Site. Astronomy lecturer Tresh Muro will be present. Space is limited. Visit these websites for updates: mostateparks.com/park/arrow-rock-state-historic-site www.facebook.com/ArrowRockStateHistoricSite

Getting There

Drive west on Interstate 70 for twenty-eight miles and take exit 98. Drive north on Highway 41 to reach Arrow Rock.

Totality Duration

2 minutes 39 seconds

Notes

Arrow Rock is directly under the eclipse centerline. For updates, check their official website at www.arrowrock.org/events.php.

Event	Time (CDT)	Altitude	Azimuth
Sunrise	6:30:00AM	0°	73°
Eclipse Start	11:44:27AM	56°	137°
Totality Start	1:10:58PM	62°	178°
Totality End	1:13:37PM	62°	179°
Eclipse End	2:38:59PM	57°	220°
Sunset	7:58:00PM	0°	285°

Battle of Lexington SHS

Elevation: 850 feet
Population: 4,726 (city)
Main road/hwy: US 24

Overview

The Battle of Lexington State Historic Site is a state-owned property situated in Lexington, Missouri. The site was preserved in 1958 in remembrance of the American Civil War, which took place in 1861. The site offers a loop trail and tours of the grounds.

The August solar eclipse will cover Lexington and the state historic site. The park is a beautiful, serene place covered with orchards and gardens. To view the solar eclipse, pack a picnic lunch and bring a lawn chair. Walk to the Anderson House and see the bullet holes in the walls. For any upcoming events on this site, visit the Missouri state park website at

mostateparks.com/park/battle-lexington-state-historic-site

Getting There

Drive east from Kansas City on Interstate 70 for thirty-eight miles and take exit 41 north. Drive thirteen miles to reach Lexington.

Totality Duration

2 minutes 27 seconds

Notes

Visit the town's website at www.visitlexingtonmo.com

Event	Time (CDT)	Altitude	Azimuth
Sunrise	6:34:00AM	0°	73°
Eclipse Start	11:42:37AM	55°	135°
Totality Start	1:09:02PM	62°	175°
Totality End	1:11:29PM	62°	176°
Eclipse End	2:37:08PM	57°	217°
Sunset	8:02:00PM	0°	286°

Mark Twain National Forest

Mark Twain
NF

Elevation:	Varies
Main road/hwy:	Varies

Overview

The Mark Twain National Forest is one of the major forested regions in Missouri that lies along the total eclipse path. The forest is divided up into several districts in the southern part of the state. Only three districts of the Mark Twain National Forest are in the path of totality: Cedar Creek, Salem-Potosi, and Fredericktown. Of these, only Cedar Creek is completely within the path of totality. The other two districts are partly in the path of totality. Only the northern section of either of these will see the total eclipse.

Getting There

See the forest website for directions: www.fs.usda.gov/main/mtnf

Totality Duration

Varies

Notes

The Mark Twain National Forest is Missouri's only national forest. There are other woodlands administered by other government agencies.

Chart times are for Womack, MO. Other location times vary.

Event	Time (CDT)	Altitude	Azimuth
Sunrise	6:21:00AM	0°	74°
Eclipse Start	11:50:26AM	59°	142°
Totality Start	1:18:21PM	64°	188°
Totality End	1:20:32PM	63°	189°
Eclipse End	2:45:54PM	55°	228°
Sunset	7:45:00PM	0°	285°

MISSOURI RIVER

Elevation:	Varies
Main road/hwy:	Varies

Missouri River
Approximate Path

OVERVIEW

The Missouri River, one of the major waterways in the United States and Missouri, will experience the total eclipse along its route. The river meanders from the western edge of the eclipse in Kansas City and travels to St. Louis. Interestingly, the total eclipse completely covers the river along its path through the entire state. Although the Mississippi River is also covered by the totality, the Missouri River experiences the eclipse for one of the longest stretches of waterways in the United States.

Anywhere along the river will be a good, cool location to watch the eclipse from on August 21. The sun will be high in the sky, so unless there are clouds or smoke, the eclipse will be completely visible.

GETTING THERE

Varies

TOTALITY DURATION

Varies

NOTES

The Missouri River travels roughly west to east across Missouri.

WASHINGTON STATE PARK

Washington SP

Elevation:	774 feet
Main road/hwy:	Hwy 21

OVERVIEW

Washington State Park has incredible Ozark overlooks. Historical buildings constructed by African-American Civilian Conservation Corps gives the park a historic sense. The 1000 Steps Trail is a 1.5-mile rugged hike through the Big River Valley.

Washington State Park lies within the path of the total eclipse. For camping information and reservations, visit the Missouri official website at

mostateparks.com/page/64196/2017-solar-eclipse-events.

GETTING THERE

Drive southeast for six miles from St. Clair Highway 30. Exit on Highway 47 and continue for twenty miles. Turn east onto Highway 21 and drive 2.5 miles to the park.

TOTALITY DURATION

2 minutes 39 seconds

NOTES

Visit the Missouri State Parks site for eclipse updates at all forty-seven Missouri state parks that will experience the eclipse:

mostateparks.com/2017Eclipse

Event	Time (CDT)	Altitude	Azimuth
Sunrise	6:23:00AM	0°	74°
Eclipse Start	11:49:13AM	58°	141°
Totality Start	1:16:37PM	63°	186°
Totality End	1:19:13PM	63°	187°
Eclipse End	2:44:22PM	55°	226°
Sunset	7:47:00PM	0°	285°

Enjoy other Sastrugi Press titles

2017 Total Eclipse State Series by Aaron Linsdau
  Sastrugi Press has published several state-specific guides to the 2017 total eclipse crossing over the United States. Check the Sastrugi Press website for the various state eclipse books: www.sastrugipress.com/eclipse/

Antarctic Tears by Aaron Linsdau
  What would make someone give up a high-paying career to ski alone across Antarctica? This inspirational true story will make readers both cheer and cry. Fighting skin-freezing temperatures, infections, and emotional breakdown, Aaron Linsdau exposes the harsh realities of the world's largest wilderness. Discover what drives someone to the brink of destruction while pursuing a dream.

Adventure One by Aaron Linsdau and Terry Williams, M.D.
  What does it take to conceptualize, plan, and enjoy your first expedition? This inspirational book contains hard-won knowledge from both authors about their experiences on expeditions around the world. The information provided in this book is useful whether you plan to climb a high peak, cross a polar plateau, or set out on a never before attempted new trek. (*Available Summer 2017*)

Lost at Windy Corner by Aaron Linsdau
  Climbing Denali is a treacherous affair. Avalanches, blinding blizzards, and crevasses have killed experienced teams. What happens when someone decides to climb the mountain solo? In this dramatic story, Aaron describes the choices made and the lessons that were learned as a result. This is more than an adventure story. It teaches defining success on your own terms in business and life. The messages will stay with you long after the end of the book.

  Visit Sastrugi Press on the web at www.sastrugipress.com to purchase the above titles in bulk. They are also available from your local bookstore or online retailers in print, e-book, or audiobook form.

Thank you for choosing Sastrugi Press.
"Turn the Page Loose"

About Aaron Linsdau

Aaron Linsdau is a polar explorer and motivational speaker. He energizes audiences with life and business lessons that stick. He delivers a message of courage by building grit and maintaining a positive attitude. Aaron teaches audiences how to eat two sticks of butter a day to achieve their goal. He shares how to build resilience to deal with constant pressure and adrenaline overload.

He holds the world record for the longest expedition in days from Hercules Inlet to the South Pole. Aaron is the second only American to complete the trip alone.

This solo expedition is more difficult than climbing Mount Everest with a team. Being alone dramatically increases the challenge. Aaron uses emotionally stirring stories to show how to overcome obstacles, impossible challenges, and unimaginable conditions. He relates these stories to business challenges and shows how the common person can achieve uncommon results.

Aaron collaborates with organizations to deliver the right message for the audience. He relates his experiences to business realities. Aaron loves inspiring audiences. Book Aaron for your next event today.

"Never Give Up"
Grit • Courage • Attitude • Perseverance • Resilience

Learn more about Aaron Linsdau at:
www.aaronlinsdau.com or www.ncexped.com.

Smartphone link

Aaron at the South Pole after 82 days alone in Antarctica.

Remember the Nebraska Total Eclipse
August 21, 2017

Who was I with? _____

_____

_____

_____

_____

What did I see? _____

_____

_____

_____

_____

_____

_____

_____

What did I feel? _____

_____

_____

_____

_____

_____

_____

_____

_____

_____

_____

_____

_____

What did the people with me think? _____

_____

_____

_____

_____

_____

_____

_____

_____

Where did I stay?_____

_____

_____

_____

_____

_____

_____

_____